The Churchill Tank

A VISUAL HISTORY OF THE BRITISH ARMY'S HEAVY INFANTRY TANK 1940-1945

by David Doyle

AF684113

Ampersand Publishing

Published by The Ampersand Publishing Group, Inc.
21 SW 1st Ave., Delray Beach, Florida 33444

Acknowledgements:
This book would not have been possible without the generous assistance of the following individuals and organizations: David Fletcher, Tom Kailbourn, Pat Ware, Scott Taylor, Don Moriarty, John Murphy, Uwe Feist, The Tank Museum, The Imperial War Museum, TACOM LCMC History Office, The Patton Museum, The Bundesarchiv and Armor Plate Press. **Cover photo:** Churchill tanks of A and B Squadrons, 43rd Battalion, Royal Tank Regiment, 33rd Brigade wait to move off as squadron leaders and tank commanders discuss operations in the foreground on 22 October 1942. (Imperial War Museum) **Rear cover photo:** A Churchill tank of 51st Royal Tank Regiment, moving through the undergrowth during an advance across the Italian countryside on 20 July 1944. (Imperial War Museum).

Sources:

Dieppe Through the Lens of the German War Photographer by Hugh G. Henry Jr. and Jean Paul Pallud. After the Battle, 1993

The Churchill Tank by Chris Ellis. Tanks Illustrated No. 25. Arms and Armour Press, 1987.

Churchill Crocodile Flamethrower by David Fletcher. Osprey, 2007.

Mr. Churchill's Tank: The British Infantry Tank Mark IV by David Fletcher. Schiffer Publishing, 1999.

British Tanks in Normandy by Ludovic Fortin. Histoire & Collections, 2005.

Churchill, Vol. 1 by Leszek Moczulski. Gunpower No. 26. AJ Press, 2008.

Tamiya Photographic Album of British Churchill Tank. Tamiya Plastic Model Co.

Churchill Tank: Vehicle History and Specifications The Tank Museum. Her Majesty's Stationery Office, 1983.

©2010, The Ampersand Publishing Group, Inc. All rights reserved. This publication may not be reproduced in part or in whole without written permission from the publisher, except in cases where quotations are needed for reviews.

Agent in the UK: Historex Agents, Wellington House, 157 Snargate Street, Dover, KENT CT17 9BZ Tel: 01304 206720 Fax: 01304 204528. E-mail: sales@historex-agents.com

Agent in Italy and Europe: tuttostoria, via G.S. Sonnino, 34-43100 Parma, Italy Tel: 39-0521-292733 Fax: 39-0521-290387.

Agent in Japan and the Far East: HobbyLink Japan, Kurohakama-cho 162-1, Sano, Tochigi 327-0813 Japan. Fax: 81-283-27-2352. E-mail: dealers@hlj.com

Table of Contents

Churchill Mk I 4
Churchill Mk II 19
Churchill Mk III 37
Churchill Mk IV 57
Churchill NA75 83
Churchill Mk V 89
Churchill Mk VI 96
Churchill Mk VII 107
Churchill Mk VIII 115

This publication is intended to visually showcase the deployment of the British Churchill tank. Rather than duplicate the fine work of other individuals, we recommend the following for those who wish to read more about the development of this unique weapon:

Mr. Churchill's Tank: The British Infantry Tank Mark IV by David Fletcher. Schiffer Publishing, 1999

Churchill Tank: Vehicle History and Specifications The Tank Museum. Her Majesty's Stationery Office, 1983

Introduction

Like its namesake, the British Prime Minister, the Churchill tank was tough, resilient, and long lasting. It was probably more adaptable than Winston Churchill was but, also like him, it had something about it that seemed to hark back to another age. And, like him, it continued to serve long past its prime.

One thing I have always liked about the Churchill was is fact that the different Marks are relatively easy to tell apart – at least I have found it reasonably easy and that enhanced my interest. Now, with this visual feast, that skill, (if that's what it is) can be extended to anyone who studies this book. David Doyle has divided the book up by Mark so that one can follow the evolution of the Churchill through from prototype to the final model. In the process, he has shown it in many varied aspects and theatres of war. He has the knack of using a caption to bring a picture to life by picking out all those fine details that we might otherwise miss.

In the history of armored warfare, particularly from the British point of view, the Churchill was a significant type. Early examples proved to be so poor that at one stage it was considered setting them up as static pillboxes around the countryside in the event of invasion—probably the worst fate you can wish on any tank. Not that it was alone. During the Second World War, across all nations, there were similar horror stories. There had been before and there have been since. But the Churchills improved as time went by and, even more importantly, their crews began to understand them and even love them. They gave them names, kept them clean, and lavished them with hours of care and maintenance. And that, of course, is the secret. You can keep a tank running, even a poor tank, if you understand it and learn to anticipate problems.

The other stimulus, inevitably, is action. Keeping a tank running is even more important when you know your life depends upon it, and that is probably the best incentive of all. Years ago, when I was writing *Mr. Churchill's Tank* for Schiffer, a friend suggested that I should call it The British Tiger. I didn't, of course—I worried that it might be misleading and that the association between the two tanks is loose at best. I can only think of two respects in which the Churchill could be described as better than Tiger. The first is quantity, of course. Churchills out-numbered Tigers many times over. The second is armor thickness, at least if you compare the Tiger with the Churchill Mark VII. Frontal armor thickness on the Tiger is about 100mm, while on late production Churchills it is 152mm, half as much again.

The Prime Minister was naturally passionate about the tank that bore his name, although some other Members of Parliament were not so keen. Indeed once, in the House of Commons, feelings got so hot that a duel was considered: Winston in a Churchill tank and his opponent in a Tiger! The result would probably have been a foregone conclusion, and there might have been a vacancy for the post of Prime Minister right in the middle of the war. In reality, of course, it never happened.

Interest in British tanks has never been that strong, not even in Britain, so I am really pleased to see this book by an American author and American publisher, and I hope that it generates a lot more interest in the subject for American readers.

—David Fletcher, Bovington, England

The Churchill tank was conceived as a tough-skinned, heavily armed tank that would support infantry during assaults. In addition to carrying a gun turret, it was designed with tracks returning at the top of the tall hull to allow room on the hull sides for gun sponsons. This is the second prototype, the A20E2, which has a superstructure fitted on the hull instead of a turret. (The first prototype, the A20E1, had a Matilda II turret.) (Warehouse Publications)

Churchill Mk I

The first production model of the Churchill, the Mk I, entered service in 1941. It was armed with a 2-pounder gun and a coaxial 7.92mm Besa machine gun in a cast turret, and a 3-inch howitzer next to the driver in the bow. Side-mounted engine air intakes are a feature of all marks of the Churchill. The Mk 1 and Mk II have rounded, cast intakes with louvers on the lower side. The tracks are the heavy BTS 3s, with an 8.29" pitch and 14" width. (Warehouse Publications)

On 23 May 1941, King George VI and other dignitaries (including the tank's namesake, Prime Minister Winston Churchill) were given a demonstration of several Churchill Mk I tanks at the test grounds of the manufacturer, Vauxhall Motors, at Luton, Bedfordshire. In this photo, the king is climbing a ladder onto the first production Churchill Mk 1, T30971. Note the cover over the muzzle of the 3-inch howitzer. (Tank Museum)

Early in the production of the Churchill Mk I and Mk II, it was discovered that the rounded air intakes tended to suck dust and debris up into the downward-facing louvers. To address this problem, a revised air intake fabricated from welded plates was developed with the intake opening at the top. These new intakes also improved the wading capabilities of the tank. Track guards (or fenders) were also developed. These guards, when installed over the upper run of tracks, helped to reduce dust. These and other modifications were made retroactively to Mk I and Mk II tanks in various rework programs. On this modified Mk II, bogie wheels with rubber tires have also been applied. (Patton Museum)

Identifying characteristics of early production Churchill Mk Is include exposed tracks all around; a cast turret with 2-pounder gun and stowage boxes on the bustle and right side (but not the left); a square side door; and rounded air intakes. With the exception of the horns (the extensions that carried the idlers and their bearings) and the riveted plates over the drive sprockets, the hull sides are quite clean. (Patton Museum)

The stowage box on the right side of the Mk I's turret held greatcoats and a machete. A single track shoe is stowed to the rear of the air intakes. In this and the following photo, the tube of the 3-inch howitzer is visible through the opening to the top rear of the idlers. Notice the design of the early-type pistol port on the turret and the eight spokes on each bogie wheel, which were later eliminated. (Patton Museum)

At the rear of the hull are two brackets for mounting an auxiliary fuel tank. The mufflers and exhaust outlets are situated on top of the engine deck. Two small fire extinguishers are visible on the deck to the rear of the mufflers. A carrier for signal flags is on the side of the turret next to the pistol port. (Patton Museum)

In the Dieppe Raid of 19 August 1942, officially known as Operation Jubilee, the British attempted to seize a German-defended port on the northern coast of France. The operation was a stunning disaster, with the raiders suffering over 50 percent casualties. An armored component of the raid, comprising the 14th Canadian Army Tank Regiment (CATR), also known as the Calgary Regiment, was equipped with 30 Churchill tanks, including Marks I, II, III, and Oke flamethrowers. A German photographer captured this image of "Burns," a Churchill Mk I of B Squadron Headquarters. Its right track was blown off as it crossed the beach. (BA 362-2209-14)

"Company," a Churchill Mk I of F Troop, 14th CATR, was commanded by Capt. George T. Valentine at Dieppe. A projectile striking the left idler disabled the tank. The side door is open and deep-wading trunks are fitted to the tops of the air intakes. Exhaust extensions are also installed. Note the red-white-red recognition flash painted on the rear of the hull. (Patton Museum)

Churchill Mk I tanks engaged in intensive testing and breaking-in exercises in the UK in the winter of 1941-42. "Indus," a Churchill Mk I of B Squadron, 9th Royal Tank Regiment, is at the lead of a column during maneuvers at Tilshead on Salisbury Plain on 31 January 1942. (Imperial War Museum)

These Churchill tanks of the 33rd Army Tank Brigade conduct exercises outside of Brockenhurst in the New Forest on 13 August 1942. In the foreground is a Churchill Mk I, identifiable by its 3-inch howitzer in the bow. Turret fronts with three vertical slots for the main gun, coaxial machine gun, and gunner's sight are also distinguishing characteristics of the Churchill Marks I and II. (Imperial War Museum)

Churchills saw action in Tunisia in 1943. In late April, these vehicles of the North Irish Horse, including a Mk I at the right, were photographed during the battle of Longstop Hill, during which Churchills provided excellent support to British infantry attempting to break through German defenses in hilly terrain. The Mk I has been fitted with the late-model air intakes and track guards, but the front mudguards have been removed. Many of the Churchills sent to Tunisia were so modified. (Imperial War Museum)

Churchill tanks participate in Exercise Spartan on 9 March 1943, with a Mk I in the lead. The vehicle has been modified with a full track guard and the later-type air inlet. The howitzer gunner is sitting in his open hatch. Tests of the Churchill Mk I revealed a number of deficiencies, including jamming of the bogies, failures of track links, blown gaskets, and corrosion of flexible lines in the lubrication system. These problems were identified and remedied during trials. (Imperial War Museum)

The Churchill CS (Close Support) was essentially an early Mk I with the 2-pounder gun moved to the bow position and the 3-inch howitzer relocated to the turret. This conversion was useful for laying down high-explosive or smoke shells, which were much less effective when the 3-inch howitzer was in the bow position. Only a few such vehicles were produced. (Tank Museum)

Top left: In this frontal shot of a surviving Churchill Mk I, the layout of the driver's vision port and the 3-inch howitzer are apparent. The hold-open catch on the door of the driver's vision port and its corresponding latch on the frontal plate are visible. **Top right:** The 3-inch howitzer is slightly elevated and the armor cover with the aperture for the gunner's sight is correspondingly tilted, exposing an opening in the hull below and behind the cover. The muzzle of the howitzer has been plugged. **Above left:** The staggered design of the frontal plate of the driver's and gunner's compartments is in view. Above the hinge of the driver's vision port is the driver's periscope mount. Also in view are the handles of the driver's and gunner's hatch doors and the ventilator cover. **Above right:** A view of the 3-inch howitzer from the roof over the gunner's compartment. Note the manner in which the armored guard is attached to a flange at the front of the gunner's sight housing, to the left of the howitzer. (Scott Taylor, all)

Top left: Although the 2-pounder gun is present, the Besa coaxial machine gun is missing from its mount to the left of the 2-pounder. The protrusion on the roof of the turret, to the left, is the port for the 2-inch smoke bomb thrower. To the right of the 2-pounder are the heads of the commander's and gunner's periscopes. **Top right:** The four small discs on the right side of the turret were tapped for bolting on a stowage box. The tracks are the heavy-type BTS 3s. **Above left:** The heavy texture and rough casting of the Churchill Mk I turret are apparent in this view. The gunner's and driver's door handles are fashioned from bent steel strips. **Above right:** As viewed from the front right corner of the turret, the opening for the 2-inch bomb thrower with its raised splash guard is to the side of the ventilator cover. (Scott Taylor, all)

Churchill Mk II

The Churchill Mk II is essentially a Mk I with the 3-inch howitzer removed and a 7.92mm Besa machine gun installed in its place. Instead of separate apertures in the frontal plate for the 3-inch howitzer and its sight, there is a single aperture with stepped sides. This example, War Department registry number T31113, bears a stencil on the side of the plate above the horn: "This vehicle is filled with anti-freeze 60-40 and must not be drained." (Patton Museum)

To stiffen the plates attached to the tops of the idler-wheel horns, as on this Churchill Mk II, triangular gussets were welded to the insides of the plates and to the frontal plate of the driver's/machine gunner's compartments. On the glacis are two headlights (an ultra-violet to the left in the photo and a blackout light to the right) and two marker lamps. Note how the shape of the aperture for the bow-mounted Besa machine accommodates the housing of the gun sight on the outboard side of the gun. (Patton Museum)

The joint between the two sections of the early-style air intake is clearly visible in this photo. The bogie wheels lack the spokes found on the earliest Mk Is. As with all Marks of the Churchill (except the later Black Prince), there are 11 sets of bogie wheels (or rollers) on each side, mounted in pairs on independent coil-spring suspensions. The two front bogies on each side are not load bearing. At the front are adjustable idlers; the drive sprockets are at the rear. (Patton Museum)

The top run of early-production Churchill Mk II tracks is fully exposed, except for mudguards at the rear. The exhaust lines and mufflers on the engine deck are also exposed, with the exception of a guard covering part of the top of each exhaust line. Access to the engine compartment is through four doors, hinged on their outboard sides, on the engine deck. To the rear of the mufflers are access doors for the transmission compartment. (Patton Museum)

Details of the turret and hull roofs are visible in this view of a Churchill Mk II. The commander's hatch is mounted on a flat, circular plate that can be rotated 360 degrees. Mounts for two wireless antennas are on the rear of the turret. Shovels are stowed on the deck over the transmission compartment, and an auxiliary fuel tank is on the rear of the vehicle. (Tank Museum)

This Churchill Mk II, registry number T31113, is viewed from the left rear. Two eyes for towing shackles are incorporated into the rear of both of the hull side plates. A somewhat battered supplementary fuel tank is mounted on the rear. The drive sprockets are virtually hidden from view by the hull side plates. (Patton Museum)

The uncluttered rear of the Churchill Mk II hull lacks a towing pintle and other accessories. The towing shackles fit very loosely in their mounting eyes. Note the small padlock for the stowage box on the turret bustle. This box held a camouflage net, waterproof cover, tent poles and pegs, rope, and strips and discs that served as ground/air communications symbols. (Patton Museum)

The box on the right side of the turret held five greatcoats and a machete. The upper run of tracks is supported by five pairs of curved steel skids on top of each sponson. The steel strip mounted above each air intake was intended to keep mud from spilling off the tracks and onto the intake, but the fixture did not work as planned. (Patton Museum)

Left: A Churchill Mk II displays its capacity for conquering obstacles. The tall profile of the tracked suspension was key to the tank's ability to cross ditches. At the rear of the engine/transmission deck is the early-type 5-inch air outlet, with separate air ducts visible within (later and reworked Churchills had an 8-inch air outlets). Two stowed shovels rest on the right side of the vent. Faintly visible to the front of the gunner's periscope is the commander's sighting stick, a crude vane sight that was eventually replaced by more sophisticated models. (Patton Museum) **Right:** "Joan," a Churchill Mk II, sits at the rear of a column of Churchills and crews awaiting inspection. The War Department registry number, T68463, is rather crudely painted on both rear mud guards, while "69," apparently the vehicle number, is also painted on the right guard. The red-white-red recognition flash is painted on the auxiliary fuel tank. The next Churchill in line is named "Jill." (Patton Museum)

The 7.92mm Besa machine guns have been removed from the coaxial and bow mounts of this Churchill Mk II, and the vehicle has been retrofitted with the full track guard set and late-type air intakes. With the installation of full track guards, the steel track skids on top of the sponsons were eliminated in favor of a pair of track guide rails. Fastened to the mud guards on either side of the driver's/bow gunner's compartment are fixtures made from bent steel rods. In the background are two Cruiser Tanks Mk V Covenanter. (Patton Museum)

"Penelope," a Churchill Mk II, leads a column of Churchills. She has been retrofitted with full track guards but still retains the early-style air intakes. The following vehicles also have the original air intakes, but lack track guards. On top of the turret is a commander's vane sight with three vertical vanes. This type of sight was common on Churchills in 1942. (Patton Museum)

This Churchill Mk II, registry number T32092, displays a Lakeman mount, a pedestal fixture for suspending a Bren machine gun for antiaircraft firing. A prominent horizontal casting seam is visible across the front of the turret. The turret of the Churchill Mk I and Mk II was a single casting with the exception of the roof, which was rolled armor plate welded or bolted to the lower casting. (Patton Museum)

The loader/wireless operator of a Churchill Mk I or Mk II waves from his hatch. A square tactical sign is painted on the area where a stowage box could be mounted. The four disc-shaped, tapped mountings for the box are visible. Note the hold-open latches on the near side of the rear door of the hatch and the far side of the front door, as well as the crash pads on the insides of the doors. (Imperial War Museum)

With the access doors and panels of the engine and transmission compartments removed, students at the Royal Armoured Corps Officer Cadet Training Unit at the Royal Military College, Sandhurst, are learning about the engine components of a Churchill. Power was supplied by the Bedford twin-six 350-horsepower engine. Approximately the first 100 Churchills had the Merritt-Brown 301C five-speed gearbox/steering system, while subsequent tanks featured the Merritt-Brown H4 four-speed transmission. (Imperial War Museum)

A Churchill Mk II exits a LCT(1) onto a muddy beach during combined-arms landing exercises at Thorness Bay on the Isle of Wight, 27 May 1942. Note the gussets at the top corners of the frontal plate of the driver's and bow gunner's compartments. A "175" unit symbol is faintly visible on the front of the bow, and a square tactical sign is on the turret. (Patton Museum)

A Churchill Mk II is deep in mud, and a tow cable is attached to the tow shackle. Details are visible for the interior of the driver's vision-port door, the bow machine gun mount, and the turret. In addition to the two small marker lamps on the glacis, two types of headlights are mounted: an ultraviolet headlight on the right side (left in the photo) and a blackout light on the left side. (Imperial War Museum)

A German military photographer captured this view of "Ringer," a Churchill Mk II of Regimental Headquarters, 14th Canadian Army Tank Regiment (Calgary Regiment) at Dieppe. It became temporarily bogged in thick chert after exiting its landing craft, and subsequently its left track was blown off. The tank's commander was Capt. A. G. Stanton. The vehicle wears a tactical symbol "Z2" within a diamond on the turret and the red-white-red recognition symbol on the side of the hull. (Imperial War Museum)

In early 1942, two Churchill Mk IIs, including this one (registry number T30999), were shipped to the Mechanisation Experimental Establishment, Middle East Branch, in the Nile Delta for trials under desert conditions. The tanks underwent various modifications, including different types of track supports and reinforced suspensions. A structure has been added to the front of the bow of this vehicle, coming almost even with the fronts of the mudguards, to divert dust from the driver's field of vision. In addition, the early-style air intakes have been turned upside down and extension trunks have been added, making the intakes even with the tops of the tracks. (Tank Museum)

Churchill Mk III

The Churchill Mk III went into production in March 1942. It is similar in layout to the Churchill Mk II, but has a larger turret of welded construction housing a 6-pounder gun. Early examples of the Mk III lack track guards and have early-type, rounded air intakes. The firm of Babcock & Wilcox Ltd. manufactured the turrets. (Tank Museum)

The layout of the front of the hull of the Churchill Mk III is similar to that of the Mk II. From the front, the new, welded turret has a slightly more massive and boxlike appearance. Between 675 and 692 Mk IIIs were manufactured, and deliveries began in late April 1942. This Mk III mounts the BTS 3 light track shoes. (Imperial War Museum)

In this overhead rear view of a Churchill Mk III, registry number T31178, the unprotected mufflers and exhaust outlets are visible on the engine deck. A Lakeman mount for a Bren machine gun appears to be lying between the hatches on the turret roof. The turret ventilator has been moved to the left side of the roof in front of the gunner's periscope, and the stowage box found on the right side of the Churchill Mks I and II has been eliminated. (Patton Museum)

"Buttercup," a Churchill Mk III, registry number T31655, was photographed on the beach in front of the Casino at Dieppe after the failed raid of 19 August 1942. The vehicle, part of 9 Troop, B Squadron, Calgary Regiment and commanded by Sgt. J. D. Morrison, laid down a carpet of chespaling—wooden slats bound together—on the beach after debarking from its landing craft. The carpet-laying apparatus was then jettisoned. The wading trunks on the air intakes were jettisoned upon landing, but the exhaust extensions remain in place. (BA 362-2207-15)

"Betty," a Churchill Mk III of 7 Troop commanded by Lt. A. L. Breithaupt, drove into the sunken entrance to an underground bunker during the Dieppe operation. The left track was shot off during efforts to retrieve the tank under fire. Bits of waterproofing material are on the 6-pounder barrel and front deck. The 7 Troop unit marking, a white 7 on a blue square, is on the turret and front of the bow. Also on the bow are the registry number, T68176.R (R standing for reworked), the name "Betty," the "175" on a blue-over-khaki square signifying the Calgary Regiment, and the Maple Leaf insignia of the 1st Canadian Army Tank Brigade. (Patton Museum)

After the end of the Dieppe Raid, German soldiers inspect "Bert," a Churchill Mk II, registry number T68560R, of 6 Troop, B Squadron, Calgary Regiment. The vehicle's left track was blown off while operating adjacent to the Casino, which is out of view to the right. The crew remained inside "Bert" and used it as a static pillbox until members of the Royal Hamilton Light Infantry assisted them in exiting the tank. (BA 362-2206-24)

Later, German engineers repaired the left track of "Bert" and used the vehicle to tow "Chief," registry number T311240R, of C Squadron Headquarters, Calgary Regiment. "Chief" was one of several Churchills fitted with jettisonable chespaling carpet layers for the Dieppe raid. The tank spent the last part of the raid emplaced on the beach as a command post. Bits of sealant material are still visible around the side door. (BA 291-1242-06)

The Germans salvaged 23 of the Churchill tanks captured in the Dieppe Raid. Some reportedly were sent to the German Army Weapons Office at Kummersdorf for technical analysis. Here, German officers view one of the captured Churchills, a Mk III. The vehicle is missing the second from the rear bogie wheel unit. Note the small fire extinguisher on the side of the turret. (Patton Museum)

In another photo from the same series, German officers and soldiers inspect a captured Churchill Mk III. The name of the tank and the War Department registry number are in white on the side of the vehicle. To the front of the door is the red-white-red recognition symbol. (Patton Museum)

In this view of the inspection of Churchills captured at Dieppe, German personnel examine two Mk III tanks. A photographer takes a picture of the turret of the tank on the left. (Patton Museum)

"Blondie," a Churchill Mk II of 7 Troop, Calgary Regiment (registry number T68880), was restored to working order and is shown here manned by a Panzer crew. During the Dieppe Raid, the left track of "Blondie" was broken, after which it served as a pillbox on Red Beach until its crew abandoned it after using all of the ammunition. Attached to the side of the track guard to the front of the driver's vision port is a wooden jack block; one was carried on either side. (Patton Museum)

This Churchill Mk III, "Feltham," was photographed during exercises in the United Kingdom on 4 October 1943. The number 5 in a triangle on the turret signifies a vehicle of 5 Troop, 1st Squadron. The Churchill in the background is towing a Rotatrailer (also see page 55). (Imperial War Museum)

Six reworked Churchill Mk III tanks were sent to Egypt, arriving in time to serve with the Special Tank Squadron (also called Kingforce) at the Second Battle of El Alamein in October 1942. Before proceeding to the front, the tanks received further modifications, including a two-color desert camouflage scheme and canvas curtains called "dodgers" draped across the front of the tank. These served to reduce the dust churned up in front of the driver's position. (Imperial War Museum)

A Churchill III tank of Kingforce was photographed in North Africa on 6 November 1942. The crewman in the foreground is removing German signal wire entangled in the tracks. Note the two-color, sand-over-green camouflage scheme. (Imperial War Museum)

Three of the Kingforce Churchill Mk III tanks appear in the desert, around the time of the Second Battle of El Alamein in October 1942. All have sand-colored camouflage schemes over the basic green of the tanks, with hard edges between the two colors. War Department registry numbers are faintly visible to the rears of the side doors. Note the two pennants on the wireless antennas. (Patton Museum)

A Churchill Mk III of the 145th Regiment, Royal Armoured Corps, accompanies infantry in an advance near Medjez-el-Bab, Tunisia on 8 May 1943. The tank has received a roughly applied sand-colored camouflage coat over its underlying green paint. A circle, evidently a tactical sign, is on the side of the bustle stowage box. (Imperial War Museum)

Churchill Mk III tanks of the 25th Army Tank Brigade, probably belonging to A Squadron, North Irish Horse, advance near Ksar Mesouar station, Hunt's Gap, Tunisia, on 28 February 1943. On the preceding day, a Churchill of the North Irish Horse commanded by Lt. W. R. Hern knocked out a Tiger tank in the area, and on the day of this photograph Churchills of that unit destroyed three PzKpfw III tanks. (Imperial War Museum)

A Churchill Mk III of the 51st Royal Tank Regiment, 25th Army Tank Brigade, appears with a Universal Carrier during an assault by the 6th Armoured Division at Pichon, Tunisia, on 8 April 1943. This regiment included 58 Churchill tanks Mks I, III and IV, as well as two Churchill ARVs. (Imperial War Museum)

A Churchill Mk III tank advances up a road near Medjez-el-Bab, Tunisia, between 23 and 25 April 1943. Note the tow cable attached to the shackle at the front center of the hull. Careful scrutiny reveals that this tank is towing a Rotatrailer (inset). This simple, unarmored trailer was designed to carry extra supplies to the edge of the combat zone. The Rotatrailer could carry 10 gallons of water, 12 days of rations, 100 rounds of 2 pdr ammunition (or 40 rounds of 6 pdr), 12 gallons of lubricating oil and 120 gallons of fuel in its spherical wheels. The trailer was employed with some success in 1941 during Operation Battle Axe, where it was towed behind Valentines and Crusaders. They were briefly deployed with Churchill units in Tunisia, but they proved very unpopular with tank crews. It was difficult to maneuver when reversing with the trailer attached, and the unit could actually lodge under the tank. (Imperial War Museum)

Churchills of the 145th Regiment, Royal Armoured Corps, of the 21st Army Tank Brigade, take up positions in a field in Tunisia in early spring, 1943. The regiment was formed in November 1941 from the 8th Battalion, Duke of Wellington's Regiment. The two nearest tanks are Mk IIIs. Swatches of a sand color have been applied over the green base color. (Patton Museum)

Churchill Mk IV

Production of the Churchill Mk IV began in mid-1942. A total of 1,622 were built. The Mk IV features a cast turret, a design that was intended to reduce the internal spalling that the welded turrets of the Churchill Mk III often suffered when hit. The chassis of the Mk IV is almost identical to that of the Mk III. The Mk IV in this photo has been reworked, as signified by the "R" suffix of its War Department number, T32418R. It mounts the short version of the 6-pounder, the Mk III. (Warehouse Publications)

The cast turret of the Churchill Mk IV is of an entirely different design from the cast turrets of the Mk I and Mk II. It is recognizable by the single aperture in the front of the turret for the 6-pounder gun, coaxial 7.92mm Besa machine gun, and gunner's sight, as opposed to the three vertical slots found on front of the Mks I and II turrets. The gunner's periscope no longer appears on the left front of the turret roof. Notice the commander's sighting stick to the front of the ventilator cover. (Patton Museum)

A Churchill Mk IV undergoes wading tests to determine if it is watertight. The tank is fitted with BTS 3 light track shoes. With the 6-pounder and Besa guns removed, details for the exteriors of the gun mounts are clearly visible. The construction of the inside of the driver's vision port door is also evident. Note the cover fitted over the opening for the 2-inch smoke bomb thrower on the turret roof, the unusual position of the ventilator near the centerline of the turret, and the absence of the driver's periscope and guard. (Warehouse Publications)

Two Churchill Mk IVs are loaded on flatcars, while cars to the side are loaded with Valentine and Matilda II tanks. The turrets of the two Churchills are traversed to the rear and the air cleaners have been removed and stowed on the engine deck, a common practice to reduce the width of the vehicle for rail transport. Visible on the nearer Churchill are a circular tactical sign on the turret and various chalk markings on the mudguards. (Tank Museum)

A close-up of the left side of "Salome," a Churchill Mk IV, provides clear details of the construction of the air intake, with a grille hinged to its top. In this black and white photo, the red sides of the recognition flash blend with the tank's green base color, leaving only the center white rectangle discernible. Note the large, conical nuts around the WD number and recognition flash. These are part of the Churchill rework program, during which the early, flush-mounted and spot-welded bolts that attached an outer layer of armor skin to the inner layer of mild steel were replaced with longer bolts and the conical nuts. (Imperial War Museum)

The front plate of the hull of "Salome" displays interesting markings. The original WD number near the bottom of the plate has faded, and the same number with a B suffix has been freshly painted above it. Above the number is a black and white bull's head with red nostrils and horn tips on a yellow background, the insignia of the 79th Armoured Division. To the opposite side of the plate is the bridge classification number, 40. (Imperial War Museum)

A Churchill tank leads a column of trucks on parade through Tunis, Tunisia, on 8 May 1943, one day after the Allied capture of the city. The nature of the hard campaigning that led up to the capture is evident by the heavy coating of dust on the vehicle. (Imperial War Museum)

A Churchill Mk IV of the 1st Canadian Armoured Division is en route to the Gothic Line on 26 August 1944. The main gun is a 6-pounder Mk IV, identifiable by the counterweight fitted at the muzzle end. "S[t.?] Teresa" is painted in the space between the driver's vision port and the bow machine gun. The door of the vision port is closed but the small, hinged shutter at the center of the door is open, exposing the vision slot. (Imperial War Museum)

"Bradford," a Churchill Mk IV, enters an unidentified village in the United Kingdom during Exercise Spartan on 9 March 1943. This offensive exercise involved 10 divisions, including four armored, and was intended to rehearse the taking of a bridgehead in occupied Europe. (Imperial War Museum)

A mud-splattered Churchill Mk IV tank rolls down a dusty trail in Tunisia on 25 April 1943. Note the counterbalance on the muzzle end of the 6-pounder main gun and the War Department number on the side of the turret. (Imperial War Museum)

A Churchill Mk IV of the 6th Guards leads several other Mk IVs through Cleve, Germany. The lead tank has both Churchill light tracks and Sherman tracks applied for extra protection. A double layer of tracks is on each side of the turret. Note the damage to the front of the right track guard and the ammunition box mounted on the side of the bustle stowage box. (Imperial War Museum)

During the Normandy Campaign, this Churchill Mk IV of the 6th Guards Tank Brigade from the 3rd Tank Battalion, Scots Guards, carries members of the Royal Scots Fusiliers on the outskirts of St-Pierre-Tarentaine. In the background, tape has been strung on posts to signify a safe route through a minefield. Already, the British were seeking to augment the Churchill's protection by attaching track shoes to vulnerable areas. (Imperial War Museum)

The turret of a Churchill Mk IV has been rotated slightly to the side and the barrel depressed, giving crewmen a better position from which to clean the barrel. The counterweight attached to the muzzle end of the gun indicates that it is a 6-pounder Mk IV. Note the commander's vane sight next to the left hand of the t-shirted crewman. (Imperial War Museum)

The barrel could also be cleaned by using a rope to pull a cleaning brush through the bore, as this crewmember of "Ivan" demonstrates. The hinged front mudguard has been raised for maintenance. Also visible are the large, conical nuts on the side of the hull, one of the components of the Churchill tank rework program. (Imperial War Museum)

A Churchill Mk IV crewman in a threadbare uniform tends to the stowed gear of his tank at a base camp. To the front of the door is the tricolor recognition sign. A small "X" appears on the edge of a track shoe to the front of the second bogie wheel. Note the covers over the muzzles of the 6-pounder and Besa machine gun. (Imperial War Museum)

Tank crewmen enjoy a cup of tea on a Churchill Mk IV. This photo may have been taken just prior to the Normandy Invasion: information written on the side door indicates that the vehicle was assigned to an LCT (3) and a placard with weight and dimensions is fitted to the commander's vane sight. The Bren machine gun in the foreground sits astride a blast shield, a protective metal sheet that was intended to protect the tops of the track guards from the blast of the main gun. (Imperial War Museum)

The commander and several other crewmen of a Churchill Mk IV named "Achilles" scan the terrain while two more Churchills advance to the front. Camouflage netting is loosely draped around the turret and side of the hull of the near vehicle, partially obscuring the WD number to the front of the air inlet. (Imperial War Museum)

Churchills are gathered in a field "somewhere in Europe" in the winter of 1944–45. The three nearest tanks all appear to be Churchill Mk IVs, with several tanks missing most of their track guards. Sherman medium tanks are visible in the background. In the foreground are several halftracks with canvas tops. (Imperial War Museum)

During the fight for Chênedollé, France, on 11 August 1944, Churchill tanks of the 3rd Tank Battalion, Scots Guards, part of the 6th Guards Tank Brigade, operate in support of the 1st Welsh Guards Infantry. A tow cable is attached to a shackle at the rear of the tank, ready for quick use. Four liquid containers are stowed on the rear mudguards, two ammunition boxes have been fastened to the upper rear of the hull, and a fuel can rack is mounted on the turret bustle stowage box. Note the Sherman track sections on the side of the turret for added protection. (Imperial War Museum)

A column of Churchills, including at least two Mk IVs in the foreground, pauses along a road. The tanks are loaded down with an abundance of Churchill and Sherman spare tracks for supplementary protection. Note the track links attached above the driver's vision port and above the opening for the 6-pounder and coaxial machine gun mount. At the rear is a captured German Panther tank. (Imperial War Museum)

Smiling civilians approach a Churchill tank in Uelzen, Germany, on 18 April 1945. Judging by their expressions and the fact that they appear to be conversing with the commander, they are probably impressed workers. The tank is most likely a Mk IV of the 6th Guards Tank Brigade. Its track guards have been removed. In addition to the usual abundance of stowed gear, two ammunition boxes are attached to the rear of the turret bustle box. On the rear of the vehicle are three German jerrycans and a bucket. (Armor Plate Press)

Two Churchill tanks of the British 3rd Division, both Mk IVs, make their way through a Dutch town during the Battle of Overloon on 14 October 1944. Although the copious track links applied to the Churchills added weight and adversely affected mobility and turret-traverse speed, the tracks did help to defeat certain types and calibers of enemy projectiles. The circular symbol on the turret of the second Churchill denotes a C Squadron vehicle. (Armor Plate Press)

Churchill NA75

In early 1944, Captain P. H. Morrell, commander of the 665th Tank Troops Workshop, Royal Electrical and Mechanical Engineers, at Bone, Algeria, developed a variant of the Churchill Mk IV that had a 75mm gun and mount from a Sherman medium tank instead of the Churchill's 6-pounder. The result was designated the Churchill Mk IV (NA75), with "NA" standing for North Africa and "75" standing for the main-gun caliber. Browning .30-caliber machine guns were substituted for the Besas in the coaxial and bow positions. The revamped tank was able to hold 81 rounds of 75mm ammunition. This photo of the vehicle is signed by a number of personnel involved in the conversion project. (Tank Museum)

Two Churchill Mk IV turrets are undergoing the NA75 conversion. The one on the left has had the opening cut for the 75mm gun shield installation, while the turret to the right has the gun and mount. Because the loader's and gunner's positions on the Sherman were the reverse of those on a Churchill, Morrell simply rotated the 75mm gun 180 degrees. (Tank Museum)

Left: Workmen lower a Sherman 75mm gun and mount into the opening of a Churchill Mk IV turret. Because of the varying angles and curvatures of the fronts of the two tanks' turrets, considerable modifications were necessary to make the 75mm gun mount work in the Churchill Mk IV turret. (Tank Museum) **Right:** A row of Churchills of the North Irish Horse provide a fire-support mission during an attack by the Jewish Brigade on German defenses in the Alfonsini-Mezzano sector of Italy on 21 March 1945. Most of the tanks are Mk IVs (NA75), but the second vehicle is a 75mm-armed Churchill Mk VII. On the nearer tank's turret, the fume-extractor vent has been moved toward the centerline of the turret for the NA75. The newly installed gunner's periscopic sight to the left of the vent is also visible. On the front corner of the turret is a signal flag holder. (Tank Museum)

A broader shot of the NA75 on the previous page. Scattered on the ground are spent shell casings and packing materials. (Imperial War Museum)

During the advance to the Po River, an NA75 crew takes a break along a road near Tanara, Italy, on 23 April 1945. In the foreground are members of the 1st London Irish Rifles. The turret is tightly wrapped with camouflage netting. (Imperial War Museum)

During the advance of the 56th Division northeast of Argenta, Italy, on 17 April 1945, a mixed group of Churchill tanks, including an NA75, roll past a cameraman of the Army Film and Photograph Unit (lower right). (Imperial War Museum)

Churchill Mk V

The Churchill Mk V is another conversion of the Mk IV. Here, the 95mm OQF Mk I tank howitzer is fitted into the turret. The vehicle was designed for close support and was particularly useful for bunker busting. It has storage for 47 fixed rounds of 95mm ammunition: a mix of smoke, high-explosive, and hollow-charge antitank rounds. A total of 241 Churchill Mk Vs were produced. (Patton Museum)

Testing of the Churchill Mk V began at Lulworth in April 1943, and the tank entered operational use in January 1944. The howitzer has a distinctive counterweight at the muzzle end with a flattened lower surface. The coaxial Besa machine gun is retained, as is the bow machine gun mount. The triangular placard between the bow gun mount and driver's vision port reads "CAUTION / UNARMOURED," indicating that this vehicle was made from unarmored steel plate for test purposes. (Imperial War Museum)

At a marshalling yard, Churchill crewmen gather around a cook fire. In addition to the Mk IV at the right, several Mk Vs are in view, including Churchills at the left, center, and right background of the photo. On the Mk V at the left, the plug of the pistol port in the side door has been opened. For extra protection the vehicle wears a mix of Churchill light tracks and Sherman steel-tread tracks on the turret and hull. (Imperial War Museum)

While the crew of a carrier stands by, a Churchill Mk V advances to the front in the Reichswald. The left track guard of the vehicle is missing. The maintenance crews who applied spare track links to Churchill tanks for supplementary protection did not hesitate to use mismatched tracks. In addition to Sherman and Churchill tracks, there is a section of German Panther track on the front deck. Notice the location of the ventilator cover at the center of the turret roof. The Churchill light track link over the driver's port is mounted loosely on a rod and was designed to swing down over the vision door when closed. (Imperial War Museum)

Troopers of the 513th Parachute Infantry Regiment, 17th Airborne Division, confer with the crew of a British Churchill Mk V of the 6th Guards Tank Brigade in Munster, Germany, on 4 April 1945, two days after the city was officially captured. This vehicle also has a single Churchill light track link hinged on a rod over the driver's vision port. (Imperial War Museum)

German POWs, including several in Fallschirmjäger smocks, pass a Churchill Mk V near the end of the war. This example is devoid of the protective spare tracks often observed on Mk Vs in combat zones. It is also missing all or most of its right track guard. (Imperial War Museum)

A column of infantry-laden Churchill tanks, including a Mk V at the center, pause on a street next to a German antitank gun. Note the missing section of track guard and the shroud over the top of the air intake on the Churchill to the left. Testing showed that loading the sides of Churchill turrets could cause strains and imbalances that slowed down the rate of traverse, a potentially deadly situation. (Imperial War Museum)

Churchill Mk VI

As viewed from above, the cast turret of a Churchill Mk VI looks much like a Churchill Mk IV's. However, the 75mm gun is a characteristic of the Mk VI. Note that the loader's hatch is not parallel to the centerline of the turret, but is rather at an angle. (Imperial War Museum)

Churchill tanks of 25th Tank Brigade advance in support of the 1st Canadian Division on 17 May 1944, the first time Churchills were employed operationally in Italy. The brigade had trained with the 1st Canadian Division at Lucera, and the two units worked well together during their assaults on the Hitler Line. The lead Churchill appears to be a Mk VI with a 75mm gun and bears the triangular markings of an A Squadron vehicle. (Imperial War Museum)

Churchill tanks of the 3rd Scots Guards, 6th Guards Tank Brigade, carry infantry from the 2nd Argyll and Sutherland Highlanders into battle near Beringe, Holland, on 22 November 1944. The lead tank appears to be a Mk VI. (Imperial War Museum)

During fighting in the Alfonsine-Mezzano sector in Italy on 14 March 1945, infantrymen of A Company, 1st Battalion, Jewish Brigade, ride on a Churchill tank probably belonging to A Squadron, North Irish Horse, the brigade's supporting armored unit. The Jewish Brigade was established in September 1944 and was comprised of Palestinian Jews. The unit fought in Italy in the final two months of the war. (Imperial War Museum)

On 2 August 1944, a column of Churchills carry infantrymen forward to the south of Caumont during Operation Bluecoat. The lead vehicle appears to be a Mk VI armed with a 75mm gun wrapped in camouflage netting, and with local camouflage applied to the turret. Blast shields are present toward the front and rear of the mudguards: they are the plates that lack ribbing. (Imperial War Museum)

A Churchill Mk VI sitting in tall weeds along a dirt road exhibits a rare example of vehicular art on a British tank: a picture of a woman that appears from close inspection to have been painted onto the front mudguard, partially covering the last several digits of the WD number. On the opposite side of the road is a Jeep and several tracked carriers hitched to towed 6-pounder antitank guns. (Patton Museum)

Churchill tanks of the 6th Guards Tank Brigade, including a Mk VI at the center, lay a smokescreen during the advance on Venraij, Holland, on 17 October 1944. Beginning in 1943, smoke generators were installed at the lower rear of the hull of many Churchills. (Imperial War Museum)

A Churchill Mk VI leads tanks of the 6th Guards Tank Brigade in an advance on 2 November 1944. A tracked carrier is smoldering to the right. (Imperial War Museum)

On 4 November 1944, a Churchill Mk VI of the 4th Grenadier Guards was photographed near Moostdijk, Holland. The turret hatches were normally left open during operations, with the exception of heavy firefights. This allowed the commander and loader to survey the situation outside. (Imperial War Museum)

A Churchill Mk VI of the 6th Guards Tank Brigade, probably from the 3rd Tank Battalion, Scots Guards, transports infantrymen of the Royal Scots Fusiliers near St-Pierre-Tarentaine, France, on 3 August 1944. (Imperial War Museum)

The crew of a Churchill Mk VI makes repairs to the tracks with assistance from engineers from the Churchill ARV Mk I parked to the rear. (Imperial War Museum)

A Churchill Mk VI of the 7th Royal Tank Regiment, 31st Tank Brigade, supports infantry troops of 8th Royal Scots during hedgerow fighting in Operation Epsom in Normandy on 28 June 1944. Bedrolls and a crate are stowed on the side of the turret. The commander and loader have their heads down inside the turret, although their hatch doors are open. (Imperial War Museum)

Two Churchill Mk VIs and a Mk IV of the 6th Guards Tank Brigade have taken up positions in Dülmen, Germany, on 30 March 1945. Note the variety of extra links festooning the vehicles. (Imperial War Museum)

Churchill Mk VII

The A22F tanks, which include the Churchill Mks VII and VIII and their variants, mark a significant advance over the A22 tanks, i.e., the Churchill Mks I through IV. The A22F was generally of stronger design, featuring an enhanced suspension and thicker hull armor of integral, welded construction instead of the composite construction (armor plate bolted over mild steel) of the earlier, A22-type hulls. As seen in this Churchill Mk VII, armed with the 75mm cannon, the A22Fs also had circular side doors and driver's vision ports, thought to be less vulnerable to direct hits. Here, Footman loops and a signal-flag holder are on the side of the turret, as well as casting marks: DU26 / EMPIRE / TP19911. (Warehouse Publications)

Visible above and to the rear of the rearmost bogie wheel is the track plough, which, when lowered, scraped mud off the tracks before it could build up under the track guards. The bolts at the center of the hull's upper rear plate can mount a coupling for a flame fuel trailer when the Mk VII was converted to a Crocodile flame thrower. To the left of the bolts is a box containing an infantry telephone and first-aid kit. Below the upper rear plate is a quick-release towing pintle, flanked on either side by smoke generators. (Imperial War Museum)

The semispherical bulges on either side of the aperture for the main and coaxial guns were intended to provide extra protection to the exposed gun mount. On the roof of the turret, the arrangement of the periscopes, 2-inch smoke discharger mount, loader's hatch, commander's cupola, and radio antenna mounts is displayed. Channel-shaped guards are fitted to the tops of the exhaust lines and mufflers. (Imperial War Museum)

A partially finished Churchill Mk VII chassis is on the Vauxhall assembly line, with a turret on a dolly next to it. For the Mk VII, Vauxhall devised a production line where the bogies of the chassis rolled on tracks, with each chassis chained to the chassis in front of it. Initially, the line of chassis was pulled along by winch power, but later, as a chassis at the front of the line was fitted with its tracks, that tank itself was used to pull the line of vehicles. (Tank Museum)

Dutch civilians wave at the crew of a Churchill Mk VII, during the liberation of Velp, Holland, on 16 April 1945. The action at Velp was the final phase of the liberation of Arnhem. (Imperial War Museum)

A Churchill Mk VII of the Scots Guards, 6th Guards Tank Brigade, crosses a bomb crater during the advance to the Elbe River on 13 April 1945, while several other Churchills wait their turn. The vehicles are carrying troops from the 10th Highland Light Infantry, 15th (Scottish) Division. The bridge is a 30-foot Churchill Bridgelayer unit. Even with its significantly up-armored turret, it was still thought necessary to add extra track-link protection. (Imperial War Museum)

A Churchill Mk VII of the 21st Army Tank Brigade passes a Churchill AVRE as it crosses the Reno River near Bastia, Italy, on 18 April 1945. In the background is a destroyed railway bridge. (Imperial War Museum)

A Churchill Mk VII of C Squadron, North Irish Horse, performs service as a battle taxi, carrying members of the 3rd Battalion, 21st Infantry (Italian) into battle near Castel Borsetti, Italy, on 2 March 1945. Note the jerrycan on the right rear mudguard. (Imperial War Museum)

Top left: The Musée Memorial, Bayeux, France, owns a Churchill Crocodile built in September 1943 and based on a Churchill Mk VII chassis. It is dubbed "Ashforder," and although the WD number T17325837 is painted on the hull, the original WD number is thought to have been T32041B. The tank was acquired from a salvage yard in Portsmouth, England, and presented to the Musée Memorial by Great Britain. The track guards, turret bustle box, mufflers and guards for the engine exhausts are missing. Note the mounting studs for spare track links to the rear of the side door. **Top right:** With the track guards missing, the track tensioner is fully visible. In the circular cutouts for the track pins on the sides of the track links, the rectangular pieces are stainless steel pin retainers that are welded in place. **Above left:** The side doors are smooth in texture, contrasting with the fairly rough texture of the hull armor. The doors have latches and operating handles on the inside only. Note the holders for a tow cable to the left and above the door. **Above right:** A flat plate has been installed where the grille was originally located, on top of the air intake. Lifting eyes are welded to the front and rear of the intake for hoisting it off its mountings when necessary. (John Murphy, all)

Top left: The hub of the drive sprocket is visible through the side of the hull, to the rear of the air cleaner. The V-shaped object to the right is the track cleaner's frame, or stripper blade, which scraped mud off the tracks to protect the mudguards (or rear fenders) from damage. **Top right:** The massive flame-fuel trailer coupling at the rear of the tank was often called "the link." To the left of its mounting hinge are the fuel line and its guard, routed below the hull. With the bustle box removed, the ventilator at the rear of the turret is visible.

Above left: In this view from the front of the Churchill Mk VII Crocodile, the center-mounted towing eye is toward the bottom. On the upper hinge strap of the circular door of the driver's vision port is a catch, with the corresponding hold-open latch visible to the left of the door. **Above right:** A close-up of the front towing eye highlights its multi-tiered welding joints. (John Murphy, all)

Churchill Mk VIII

The Churchill Mk VIII was essentially the Mk VII with a 95mm howitzer replacing the 75mm main gun. The howitzer is fitted with a counterweight at the muzzle end with a flattened lower surface, like the 95mm howitzer in the Churchill Mk V. (Patton Museum)

Like the Churchill Mk VII, the Mk VIII lacks pistol ports in the turret and the side doors. Both marks were also upgraded with the more powerful Merritt-Brown four-speed transmission with controlled differential and steering. On the roof of the turret is the commander's vane sight No. 2 Mk I, which comprised a front and rear vane assembly secured with braces, giving the sight a cage-like appearance. (Patton Museum)

In this view of the left side of the same Mk VIII, the No. 2 Mk I vane sight is visible in profile. At the rear of the mudguard, the side frame of the track cleaner is in view. Casting marks appear on the side of the turret. The mudguards sweep up at a slight angle at both ends of the vehicle. (Patton Museum)

The blades of the track cleaners are visible below the bottoms of the rear mudguards.On the rear of the hull, holes are tapped and plugged for installing a Crocodile flamethrower coupling if necessary. During production of the Churchill Mks VII and VIII, these provisions were introduced on all vehicles leaving the production lines. (Patton Museum)

Features at the rear of a Churchill Mk VIII are on display. The cupola appears to be the No. 1 Mk II type, which allows the two doors to lie horizontally when open. Fire extinguishers are in brackets on the sides of the turret bustle box. Also visible are the plugged mountings for a Crocodile flamethrower kit, infantry telephone/first-aid kit, quick-release towing pintle, track cleaner blades, and smoke generators to either side of the towing pintle. (Patton Museum)

The track guards of the A22F tanks are similar to those of the A22s, with ribs on the top and a circular recess in the central portion of both guards to allow clearance for the rotation of the turret. The counterweight of the 95mm howitzer is visible to the front of the turret, and the No. 2 Mk I vane sight is seen from a different angle. Notice the guards on top of the mufflers and exhaust lines. (Patton Museum)